Miracle Auto Battery

A Deep-Cycle Battery for the Twenty-First Century

Frank Earl

iUniverse, Inc.
Bloomington

Miracle Auto Battery
A Deep-Cycle Battery for the Twenty-First Century

iUniverse books may be ordered through booksellers or by contacting:

iUniverse
1663 Liberty Drive
Bloomington, IN 47403
www.iuniverse.com
1-800-Authors (1-800-288-4677)

Because of the dynamic nature of the Internet, any web addresses or links contained in this book may have changed since publication and may no longer be valid.

Any people depicted in stock imagery provided by Thinkstock are models, and such images are being used for illustrative purposes only.

Certain stock imagery © Thinkstock.

ISBN: 978-1-4620-0533-8 (sc)
ISBN: 978-1-4620-0534-5 (ebk)

Printed in the United States of America

iUniverse rev. date: 06/23/2011

Contents

Chapter 1

Chapter 2

Chapter 3

Chapter 4

Chapter 5

Chapter 6

Chapter 7

Chapter 8

Chapter 9

SOURCES OF RESEARCH MATERIALS

PEREZ, R. A. (1985) The Complete Battery Book, Blue Ridge Summit, PA. TAB Book, Inc.

Weir, W.J. (1987) Electronic Circuit Fundamentals, Englewood Cliffs, NJ. Prentice-Hall, Inc.

Leitman, S. and Brant, B. (1994) Build Your Own Electric Vehicle, New York. McGraw Hill.

Armstrong, R.E., (1986) Fundamentals of Direct Current, Blue Ridge Summit, PA. TAB Book, Inc.

Smith, R. E. (2007) Electricity (for Refrigeration and Heating and AC), Clifton Park, NY. Delmar

Sometimes innovation comes from completing or expanding on the work of others where they have left off. I compiled data from several sources to develop a logical and scientific approach to a mechanical, rather than a chemical, solution to the need for a more efficient and longer-lasting battery for electric and hybrid vehicles.

Richard A. Perez, the author of *The Complete Battery Book*, made reference over two decades ago to the idea of charging the battery at will as an alternative to the old cycling processes that we have been using with the automotive (lead acid) battery since the early nineteenth century. I can't say for sure that the idea of charging the battery at will was an idea that was brewing in his head at that time or whether it was just a suggestion to illustrate a point he was making in his book. Maybe the idea of charging the battery at will was overshadowed by the fact that it would be impossible to simultaneously charge and discharge a deep-cycle automotive battery on the same set of terminals.

Reading the literature of others on the cycling functions of the deep-cycle automotive battery and capacitors, I realized that we could change the mechanical structure of the automotive battery so it has more than one entrance through which electrons could travel, like that of the capacitor. Perez's idea of charging the battery at will for a more efficient battery cycling process for the deep-cycle automotive battery is possible.

Acknowledgments

Some may ask what qualifies me to write this book. I suppose what qualifies me is the two years I spent reading the literature of those who came before me in the field of battery technology, who gathered all the necessary information a layperson like me needed to understand the mechanics of the lead acid battery. I took that information and expanded on it by compiling and comparing it to come to a logical and scientific approach to a more efficient battery cycling process.

Neither Allesandro Volta nor Pieter Van Musschenbroek of Leydan, Holland, was the forerunner of their discoveries, the battery and the capacitor; there were others before them who led the way with their initial experiments or discoveries, such as Von Kleist with the Leyden jar capacitor in 1745 or Luigi Galvani's 1786 phenomenon with the frog. Volta expanded on this phenomenon, and then came the battery or Voltaic-pile. My point here is that we learn by trial and error, or we learn from others. In my case, I did both.

I wish to acknowledge the support and encouragement of relatives who supported my wife and me during this long project. I offer special thanks to my brother-in-law, Derrick Watkins, a Master electrician: Prof. Gary J. Hayward and Jon R. Eddy, HVACs technician instructors, and others who encouraged me to go on with my concept and design for a modified version of the deep-cycle battery for the twenty-first century.

PREFACE

If we could charge the deep-cycle battery at will, it is possible that we could revolutionize the electric vehicle and end our dependence on foreign oil forever. I am not saying that I have the best solution to the problem of how to charge the deep-cycle battery at will, but that we must consider it as an alternative to a chemical solution. We have searched for decades to find that perfect chemical composition that would give us a longer-lasting battery for the electric vehicle, and we haven't found one yet. Maybe it's time to look at a mechanical solution.

If you think that charging the battery at will is science fiction at best, then ask yourself these questions: why is it possible to discharge and recharge the same battery hundreds of times? Is it by accident or by design? I hope this book will open the door to a conversation about a mechanical solution, rather than a chemical solution, to the need for a longer-lasting battery for the electric vehicles that scientists have been talking about for decades. Engineers are just beginning to apply the concept of regenerative braking, a mechanical alternative to a chemical solution, to extend the range of electric and hybrid vehicles on battery power.

This book brings together the scientific facts about the cycling functions of the deep-cycle (lead acid) battery based on an innovative mechanical approach to the development of a more efficient and longer-lasting battery for electric and hybrid vehicles. The answer to this problem has eluded the field of automotive battery technology for decades.

INTRODUCTION

Charging the battery at will refers to adding electrons back to the battery without interrupting the electron flow to a load source. Even though the concept of charging the battery at will is similar to the cycling functions of the run capacitor in an Ac circuit, you will probably not hear the phrase "charging the battery at will" in the field of automotive battery technology related to the cycling functions of the deep-cycle batteries for electric and hybrid vehicles today. You probably won't hear reference to "simultaneously charging and discharging the battery" either in the modern world of automotive battery technology. Charging the battery at will and simultaneously charging and discharging the battery mean the same thing, except that charging the battery at will refers to the capability of the battery, and simultaneous charging and discharging the battery refers to the process of adding and subtracting electrons from the battery at the same time. I will be using these two phrases throughout this book to explain the concept and the design of the Miracle Auto Battery.

After the chemical energy of the secondary cell battery has been used up, whether the battery's chemical energy can be replenished is the fundamental difference between primary cell (non-rechargeable) and secondary cell (rechargeable) batteries. Those in the field of automotive battery technology have forgotten this basic fact about the difference between the two types of batteries. As a result, they have been on a crusade for decades to find a chemical solution to a longer-lasting battery when all the time another kind of solution has been staring them in the face.

The fundamental difference between a secondary cell battery and a primary cell battery is the ability to replenish its chemical energy with electrical energy. The solution to a longer-lasting battery for the electric vehicle should be simple, if we redesign the deep-cycle battery so that we can replenish its chemical energy while using its chemical energy.

The information in this book is an effort to remind those in the field of automotive battery technology that a chemical solution is not the only solution to a longer-lasting battery for electric and hybrid vehicles. There is an alternative solution, just as there was for the primary cell battery when Gaston Plante combined a chemical and a mechanical solution to create the secondary cell (rechargeable) battery. All we need to do now is expand on Plante's mechanical solution so that the secondary cell battery can be used as a constant source of power for electric vehicles.

CHAPTER 1

CHEMICAL COMPOSITION

For decades, those in the field of automotive battery technology have been searching for the perfect chemical composition that would give the deep-cycle automotive battery a longer-lasting chemical reaction that could power the electric vehicle beyond commute status to become our primary source of transportation.

Over two centuries have passed since the first battery was made by Allesandro Volta, but the closest we have come since then to a perfect chemical composition is the lithium-ion crystal battery, whose cell voltage rating is just a little higher than that of the lead acid battery cell and which we still have to recharge regularly, just like the lead acid battery.

Many in the field of automotive battery technology think that the solution to the problem of the deep-cycle battery's potential to power the electric vehicle beyond commute status is to find a perfect chemical composition, but I think that they have misdiagnosed the problem for decades. Since the chemical energy of the deep-cycle battery can be replenished hundreds of times, the problem with the deep-cycle battery is not the chemical composition but the method used to replenish its chemical energy.

If you had a serious oil leak from your engine, you wouldn't be looking for a different brand of motor oil to fix the leak; you would be looking for a way to plug the leak. You would know you needed a mechanical solution, not a chemical solution. The problem with the deep-cycle battery is a mechanical problem that concerns its two-terminal design, not its chemical composition. In order to solve this problem, we must change the design of the deep-cycle battery so it has a more efficient battery cycling process.

Even after decades of using the deep-cycle (lead acid) battery to power our electrical devices, we have not yet realized that the deep-cycle battery's potential to power the electric vehicle beyond commute status is limited by its two-terminal design. It doesn't matter whether you use a lead acid battery or a lithium-ion battery; eventually it will have to be recharged if you are going to keep using it. My point is this: why spend decades searching for a chemical composition that we have to recharge with electrical energy anyway?

Scientists say that the only endless source of energy is nuclear fusion, but we could be centuries away from developing a nuclear fusion process that could be used in our automobiles. After all, battery technology has been around for two centuries, and in all that time, while it has progressed from the basic voltaic-cell battery to the advanced lithium-ion battery, we still can't power the electric vehicle beyond commute status without having to use fossil fuel or wait around for hours to recharge.

THE SAME OLD DESIGN

If we are going to revolutionize the electric vehicle and end our dependence on foreign oil, those in the field of automotive battery technology must take an in-depth look at the mechanical structure of the deep-cycle battery and evaluate the effect its mechanical structure (two-terminal design) has on its efficiency and overall capacity. They have been using the same old basic design (two terminals) for the automotive battery for more than a century. Did anyone stop to think that maybe, when it came to the deep-cycle battery, the two-terminal design was not as good a concept as it was for the automotive starting battery? They increased the thickness of the plates by adding more lead antimony when they designed the deep-cycle battery for longevity, but other than that, the basic automotive battery design hasn't changed much over the last hundred years or so.

The purpose of the automotive starting battery is markedly different from that of the deep-cycle battery, yet the deep-cycle battery is patterned after the starting battery. The automotive starting battery was designed with one purpose in mind: to start the engines of our gas-guzzling automobiles. However, the deep-cycle battery was designed to be a constant source of power.

The designers made changes so it could be deep-cycled, but they failed to realize that its two-terminal design was too inefficient for the deep-cycle battery because it has to stop one cycling process to start the other.

The two-terminal design works just fine for the automotive starting battery because it is not under a constant load, as the deep-cycle battery is. Why in the world would we try to use the same old two-terminal designs for the deep-cycle battery that we use for the starting battery? It is like using a primary cell battery instead of a secondary cell battery to start your automobile's engine, and after several starts, we just throw it away and buy a new one. My point is that the primary cell battery wouldn't be an idea battery to use on a day-to-day basis to start our engine because we can't recharge it *after* using it. The same holds true for the deep-cycle battery. Because of its two-terminal design, it is not an idea battery for the electric vehicles because you can't replenish its chemical energy while using it. Scientists should be asking these questions: Is there an alternative solution to a chemical solution for a longer-lasting battery that would help increase the overall range of the electric vehicle on battery power? How much time are we going to spend searching for a perfect chemical composition before the oil in the earth runs out?

REGENERATIVE BRAKING

There is some good news lately that engineers have been developing a mechanical solution, called regenerative braking, to a longer-lasting battery that helps extend the range of the electric and hybrid vehicles on battery power by adding electrical energy back to the battery when the vehicle's brakes are applied.

Now, regenerative braking is a comparatively new concept in which, when the driver applies the brakes, an electrical charge goes back to the batteries to recharge them, extending the travel range of the vehicle on battery power. *Charging your batteries at will* is a step beyond regenerative braking. With regenerative braking, you have to apply your brakes to send an electrical charge back to your batteries to recharge them, so this might work well in city driving, but not so well on the open road. (You would cause some concern if you were driving on the freeway, applying your brakes randomly to charge your battery.) It's also the case that whatever energy is needed to apply your brakes will go to your brakes first, and only the energy that is left over will be applied to recharging your batteries. (Remember, safety comes first.) In other words, when you apply your brakes, you might be recharging your batteries and you might not.

Charging at will is a mechanical concept as well: It allows you to send an electrical charge to your batteries at will. You don't have to stop using your batteries or apply your brakes, and you don't have to worry about whether there will be enough electrical charge to send back to your batteries to extend the range of the vehicle. I believe that, if we redesign the deep-cycle battery so we can send an electrical charge back to our batteries at will, we can go beyond the concept of regenerative braking to a longer-lasting battery for electric and hybrid vehicles.

Think of our cell phones today as an example; we can make and receive calls while charging them. Unlike the cell phone, though, electric vehicles can't run off battery power while they are recharging their batteries. Why? Because the charge and the discharge process of the deep-cycle battery is performed on the same set of terminals. Unlike that of the deep-cycle battery, the modern day cell phone batteries have separate terminals for charging and discharging; therefore, the charging process of the cell phone's battery doesn't interrupt the electron flow to the load source during its charging process, because the charging and discharging processes are carried out on separate terminal connections. The cycling process that we are using for the deep-cycle battery today is primitive and inconvenient compared to the battery cycling process that we are using for our other electronic devices—cell phones, iPods, computers, and so forth. We must think along the same lines with the deep-cycle battery.

During the deep-cycle battery's cycling process, it has either a load on it or a charge on it, but not both, and this is where the problem lies. Yes, we have the electronic alternator regulating control system that monitors the voltage levels in our automotive starting batteries today, but there is not a load and a charge on the battery at the same time. The voltage regulator system monitors the voltage level in the battery, and when the voltage level in the battery drops below a set level, the voltage regulator allows the alternator system to recharge the battery until it reaches its set level again. There is no load on the battery during the charging process; the alternator carries the electrical load of the car while the engine is running.

The cycling process used for the automotive starting battery doesn't work for the deep-cycle battery because the deep-cycle battery is the primary source of power for the load source; if we interrupt the power supply to the load source to recharge its battery, the device doesn't work anymore, or if we tried to simultaneously charge and discharge the battery on the same set of terminals, we can't charge the battery properly. Since it is not beneficial to simultaneously charge and discharge the battery on the same set of terminals, we have to stop one cycling process to start the other. This limits the deep-cycle battery's potential to power the electric vehicle so it can become our primary source of transportation.

If the deep-cycle battery is going to power the electric vehicle beyond commute status, we must think outside the box (battery) when it comes to its design. We can't design the deep-cycle battery based on the two-terminal concept of the automotive starting battery, because the two batteries have two very different purposes. Remember that the automotive starting battery is designed with one purpose in mind: to start up our gas combustion engine. Then it sits there under constant charge by the alternator regulator control system. When we use the automotive starting battery to start our engines, most of the time we use less than 1 percent of its stored capacity to do so.

The purpose of the deep-cycle battery is to be a continuous source of power for a load source, not a temporary source of power. The two-terminal design of the deep-cycle battery creates an inefficient cycling process that limits the battery's potential to power the electric vehicle beyond commute status. This is one of the major reasons that the electric vehicle has not been revolutionized: you can't travel long distances without having to stop and recharge your batteries for several hours or worrying about having enough power to make it from point A to point B or finding a suitable place to recharge your batteries when you get there.

CHAPTER 2

CHARGING THE BATTERY AT WILL

At the beginning of the Twenty-First Century, the general population just begun to hear about the concept of regenerative braking, a mechanical solution. While reading Perez's, *The Complete Battery Book*. I realized that Perez had referred to a mechanical solution, rather than a chemical solution, to a more efficient battery cycling process for the deep-cycle (lead acid) battery more than two decades ago.

Perez was talking about charging the battery at will. His assumption was that, if we could charge the deep-cycle (lead acid) battery at will, we could use a voltage limitation device to complete the charging process when the battery is almost full and use the current limitation device to start the charging process when the battery is almost empty (p. 32–33).

Perez referred again to the idea of charging the battery at will when he was explaining how the depth of discharge of the deep-cycle (lead acid) battery plays a role in the overall efficiency of the battery. "If we had the ability to fill the battery at will," Perez said, "then a depth of 80 percent is the most efficient overall for the deep-cycle (lead acid) batteries." He was referring to the amount of energy that we could remove from the batteries during their discharge cycles. In short, Perez was talking about adding electrons back to the deep-cycle battery without interfering with its discharge process to a load source as a more efficient battery cycling process.

At first, I didn't understand what Perez meant by charging the battery at will, but I had a mental picture of a battery with a load and a charging device connected across the terminals, and I said to myself, "This is no different from connecting two batteries together in series or parallel. The charging current will not enter the battery if there is a load and a charge on it at the same time. The current would follow the path of least resistance—that is, directly toward the load source—and not into the battery because the resistance in the battery is greater than the resistance toward the load source. Therefore, you can't recharge the battery while having a load on it at the same time."

Saying to myself over and over what Perez meant by charging the battery at will, I wondered if it was possible to do both the charge and discharge process at the same time on the same set of terminals. After months of reading about the mechanics of the deep-cycle (lead acid) battery, I read a book by Walter J. Weir, *Electronic Circuit Fundamentals*, in which there was a chapter on capacitors. About halfway through the chapter, I began to understand how it may be possible to expand on Perez's conjecture about charging the battery at will.

Perez's assumption about charging the battery at will as a more efficient battery cycling process for the deep-cycle (lead acid) battery became clear to me that day when I compared the mechanical structure and the cycling functions of the run capacitor with those of the deep-cycle (lead acid) battery: two different devices similar in their cycling functions of the electron current. I formed a theory based on the run capacitors simultaneous charging and discharging process in an Ac circuit. I thought that if we added additional terminals, straps, and tab connectors to the battery on the opposite side of its plates and cells, we could use the mechanics of the direct current to charge and discharge the battery simultaneously, like the alternating current does to the run capacitor in an Ac circuit without interrupting the current flow to the load source.

I began to think of charging the battery at will—simultaneous charging and discharging of the battery—like using a garden hose to water a lawn. The garden hose connects a water supply outlet valve and a sprinkler system. When you turn on the water to water your lawn, the water fills the entire length and diameter of the hose before it has enough pressure to force itself out of the sprinkler system and onto the lawn. Once the garden hose is filled with water, there is a constant water pressure between the water outlet valve and the sprinkler system, even though water is flowing out of the sprinkler system at the same time. The garden hose remains filled with water because water is being added and subtracted from the garden hose at the same time and at the same rate. This is what the concept of simultaneous charging and discharging the battery means: we are adding and subtracting electrons from the battery at the same time.

The terms "simultaneous charging and discharging" and "charging the battery at will" may not be used in the world of automotive battery technology as applied to the cycling functions of the deep-cycle battery for electric and hybrid vehicles, so I define them as they relate to my concept and the design of the Miracle Auto Battery. Charging the battery at will refers to the ability to recharge the battery without interrupting its discharging process to a load source, so you can send an electrical charge back to your batteries at the same time that you use them to power your electric vehicle. Charging the battery at will is part of a mechanical solution to a longer-lasting battery, like regenerative braking is, except you don't have to apply your brakes or interrupt the current flow to the load source to do so.

MY CRUDE EXPERIMENT

To test my theory that the lead acid battery could be charged at will if it could be charged or discharged on separate sets of terminals during its cycling processes, similar to that of the run capacitor. I purchased several used starting batteries and deep-cycle batteries. In my garage, I cut open a couple of batteries whose plates and terminals were intact to see how the battery was made and to determine how I could add additional terminals to the battery to test my theory. It seemed impossible at first because the lead-peroxide and sponge-lead plates were so fragile; especially when the battery had been used a lot, they just fell apart at a touch.

I came up with an idea to see how the battery would respond with a charge and a load on it at the same time using a 16-volt, 40-amp charging device, a 12-volt tire inflation device, and a 12-volt, 200-amp starting battery. First, I placed the charging device across the terminals of the battery, with part of battery's top casing cut off to see how the electrolyte solution would respond during the charging process. The electrolyte solution closest to the terminal connections started bubbling, like water brought to its boiling point on the stove, and I smelled gases being released.

I watched this process for about two minutes before putting a load on the battery to see how the electrolyte solution would respond during the discharge process. As the tire inflator was running, the electrolyte solution started bubbling around the terminal connections, just as it had during the charging process. Now it was time to test my theory to see whether it is possible to have a load and a charge on the battery at the same time.

In each cell of the battery, there is a negative and a positive strap that connects the plates to their respective plates in parallel (see figure T-4). I put the charging device across the terminal post without turning it on and then connected the tire inflation device to the negative and positive straps across from each terminal connection of the battery. The tire inflation device started to run as I made the connections to the negative and positive straps, although it was not connected to the terminals but to the opposing strap in each cell of the terminal connections. So I had a negative and a positive connection on both ends of the battery (see figure T-2).

For safety reasons, I attached a twenty-five-foot extension cord to the charging device so I could plug it in from a safe distance (in case the battery exploded when I turned the charger on). I stood outside my garage, behind the door, about twenty feet away and then plugged in the charger. I heard the charger turn on. There were no explosions or sparks, but I waited a couple of minutes before I went back into the garage to see what was going on. I could still hear the tire inflation device and the charger running at the same time, a good sign. When I finally went back into the garage, I could see that the electrolyte solution was bubbling and boiling around the tire inflation device connections and around the terminals where the charger was connected. Yes! It is possible to charge the battery at will. If we use separate sets of terminals on the same battery—one set for the charging current and one set for the discharging current—my crude experiment revealed that we don't have to charge and discharge the battery on the same set of terminals in order to recycle it as we have been led to believe for decades. It doesn't matter if the charging current is reversed through the same terminals or come from a separate set of terminals: as long as, the charging current comes in contact with the plates, we can reverse the chemical reaction of the battery.

Our ability to recharge the lead acid battery has nothing to do with the process of reversing the current in the charging process. Reversing the current flow through the terminals is the only way we can recharge the battery, because there is only one entrance for the current to travel in or out of the battery, so, naturally, you would have to reverse the process in order to recharge the battery. Reversing the chemical reaction of the lead acid battery doesn't have to be done on the same set of terminals; to break the bonds between the lead and sulfate ions, just charging the battery with a higher voltage than its normal battery voltage (electrolysis) should do the trick.

After two years of research on the deep-cycle (lead acid) battery's cycling functions, it was clear to me that Perez's conjecture can become our reality: the ability to replenish the battery's chemical energy while using its chemical energy, charging at will.

CHAPTER 3

THE BATTERY AND THE CAPACITOR

The battery and the capacitor were both discovered in the eighteenth century by two different scientists and years apart. Batteries are better known than capacitors are because the capacitor is hidden inside most electronic components.

The capacitor grew parallel to the battery as a source of capacitance in our electronic devices known as the condenser, which had been developed by Professor Musschenbroek of Leydan, Holland. The capacitor's applications are practically unknown to the average person today, even though the capacitor behaves somewhat like the battery in that they can both be charged and discharged, and they both store electrical energy. But their main differences, other than their shape and size and the different electromotive forces they use to move electrical current, were the differences in their cycling processes.

The battery has come a long way since the voltaic-cell battery, a primary cell battery. Now we have the secondary cell battery, also known as the storage battery or the automotive deep-cycle (lead acid) battery. Gaston Plante constructed the first secondary cell battery (rechargeable) in 1859, more than fifty years after Volta made the first primary cell battery (non-rechargeable). This type of secondary cell battery is primarily used in cars today, over a hundred and fifty years later.

The capacitor has come a long way as well. It was once known as a condenser, a device with some kind of fluid in it, because the early experimenters believed that electrical "fluid" was condensed in the Leyden jar. Today's capacitors are made with solid materials, not liquids, called dielectric plates.

If you place the capacitor and the automotive battery on the same scale to evaluate their designs and cycling functions, even though they both were developed in the eighteenth century, the capacitor has clearly evolved farther and faster than the automotive battery. We have been designing the automotive battery with the same two-terminal system and using the same cycling functions since the nineteenth century. Unlike that of the capacitor, the charge and discharge process in the battery has to be performed on the same set of terminals.

The capacitor has evolved over the last two centuries so its charge and discharge process can be carried out on multiple terminals. The automotive battery has evolved over the last century in terms of its appearance and the chemical compositions used to increase its efficiency and overall capacity, but scientists and engineers have not looked sufficiently at its two-terminal design to see what effects it has on the battery's efficiency and overall capacity. My point here is this: the automotive battery's mechanical structure hasn't changed to facilitate the uses that are needed in the modern world of electronic technology.

When the automotive battery was designed to start the automobile engines in the early 1900s, it was suited to the job, but not to the job of providing a constant source of power to electric vehicles.

The differences that I noticed between the deep-cycle (lead acid) battery and the run capacitor (a non-polarized electrolytic capacitor) other than their cycling functions were their internal design. The run capacitor is designed to conform to the mechanics of the alternating current, but the deep-cycle battery is not designed to conform to the mechanics of the direct current. As a result, we have to reverse the current flow through the battery in order to recycle it. Thus, the deep-cycle battery has only one way for current to enter or exit the battery during its cycling processes, while the run capacitor has more than one way.

After months of compiling data on the mechanics of the lead acid battery based on others' research, I came to the conclusion that the ability to charge the battery at will is neither complex nor impossible and that it can be done through the design of the battery, rather than by the mechanics of the current alone. The question is, how we can design the battery so we can add electrical energy back to it at will without having to stop the discharge cycle to start the charging cycle? I believe that this will be a simple task if we focus on the mechanical structure of the deep-cycle battery, not on its chemical structure as we have done for decades.

THE GLASS CAPACITOR

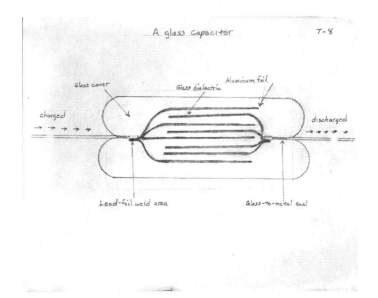

Let's look at the capacitor, especially the run capacitor, so we can see why it will be possible to charge the deep-cycle (lead acid) battery at will. The Miracle Auto Battery's concept and design is inspired by the run capacitor design and cycling functions; after all, they are cousin. The run capacitor gave me insight into how to add electrical energy back to the battery without interrupting the battery's discharging process to the load source: having separate charging and discharging terminals so current can enter and exit the battery at the same time.

A run capacitor has two or more terminals connected to individual plates in the capacitor: its plates are separated by non-conductive material called dielectric. As a result, the run capacitor can be charged or discharged on either terminal because of its separate plates and terminal connections. In an Ac circuit, a run capacitor is charged on one plate at the same time, it is discharged on the other back through the circuit, because of the mechanics of the alternating current in the circuit. Thus, the run capacitor is simultaneously charged and discharged in the circuit.

What makes the deep-cycle (lead acid) battery a good candidate for simultaneous charging and discharging is that we use a direct current to add electrical energy back to the battery and we use a higher voltage than the normal battery voltage to do so; therefore, we can simultaneously charge and discharge the battery by forcing electrons in and out of the battery at the same time using the higher voltage pressure from the charging source, similar to the run capacitors cycling functions in an Ac circuit, where the alternating current simultaneously charge and discharge the run capacitor.

Think of simultaneously charging and discharging the battery as an equalizing charge; it is not an overcharging process, because we are discharging the battery at the same time. Heat from the charging cycle is minimized because there will be less heat from the charging cycle because the battery has less internal resistance as we continue to charge it while discharging it. In short, we minimize the amount of heat and energy loss by adding and subtracting electrons at the same time.

By looking at the glass capacitor in figure T-8, you can see its plates and terminals are separated by glass, a dielectric material. Unlike capacitors, the lead acid battery's plates and terminals are all interconnected in some form or fashion: either by straps or by the electrolyte solution. Figure T-4 shows that the negative and the positive plates in each cell of the battery are not in contact with one another physically, but they are in contact by the electrolyte solution. This is one more reason; the lead acid battery is a good candidate for simultaneously charging and discharging.

Unlike the alternating current that change directions in an Ac circuit, direct current can flow only in one direction in a circuit. This one directional flow, tells us that it is possible to simultaneously charge and discharge a battery on separate sets of terminals, if the charging and discharging terminals are interconnected by way of their respective plates. Thus, the charging current can enter one negative terminal at the same time; current is exiting the other negative terminal to the load source.

Figure C-1, which shows run and start capacitors, and figure C-2 shows a picture of the traditional deep-cycle battery with a two terminal design. The reason run capacitors have more than one set of terminals and more than one way for current to travel in or out at the same time, because of its separate plates and terminal connections and its plates are non-polarized (neutral). Thus, the run capacitor can be charged or discharged on either terminal connection. Unlike the run capacitor's plates, the deep-cycle battery's plates are polarized (negative and positive); therefore, current can only enter or exit the negative terminal during its cycling processes.

FIGURE C-1

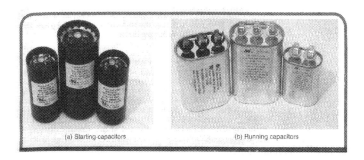

(a) Starting capacitors   (b) Running capacitors

The capacitors have more than one set of terminals and more than one entrance.

FIGURE C-2

The deep-cycle batteries have only one set of terminals and one entrance.

We use start and run capacitors to help start and run the induction motors in our electrical devices. These capacitors are installed with different types of motors—a capacitor start run motor, permanent split capacitor motors, and other types of induction motors—to help them start and run efficiently. The capacitor behaves like the battery in that it can be charged and discharged and it stores electrical energy. At its core, the capacitor is a miniature battery, so what it all boils down to is how the current is recycled through the devices. Once the start and run capacitors are installed with the induction motors, we don't have to worry about having to stop its discharge cycle to start its charge cycle, even though the capacitor is recharged after it is discharged through an Ac circuit.

What makes the run capacitor's cycling function more efficient than that of the deep-cycle battery is that, the run capacitor is designed to conform to the mechanics of the alternating current in the circuit. Designing the deep-cycle battery according to the rules of direct current, is the key to a more efficient and longer-lasting battery for the electric vehicles. When more terminals are added to the deep-cycle battery, the battery's cycling functions can interact with one another like those of the run capacitor: as electrons leave the battery to the load source, electrons are added back to the battery by the charging source, creating a more efficient battery cycling process for the deep-cycle battery.

CHAPTER 4

THINKING OUTSIDE THE BOX (BATTERY)

We must change the way we perceive the deep-cycle battery if we intend to power the electric vehicle beyond commute status. Since Volta made the first battery centuries ago, we have perceived the battery as a one-directional device based on the old water jug analogy (see figure R-7), which describes the voltage pressure in the battery as water being poured from a jug. The more water that is poured out of the jug, the less pressure will be available to push out the remaining water. The same scenario applies with the deep-cycle battery: as more voltage is used up without being replenished, less voltage is available to push out the remaining current in the battery. Like the water jug, the deep-cycle battery is one-directional. You can add or subtract electrons from the battery, but you can't do both at the same time because the battery has only one entrance through which the electrons can travel in or out.

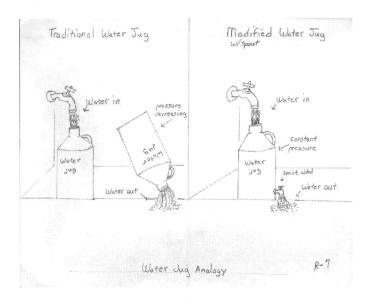

The water jug analogy concept works fine for the automotive starting battery because the starting battery is not under a constant load and you can add electrons back to the battery to keep it charged when you are not using it. However, if you want to rely on the battery as a constant source of power, the concept of the water jug can't be used for the battery. This is what this book is all about: the mechanical structure of the deep-cycle battery, not its chemical structure. If the deep-cycle battery is going to power electric vehicles beyond commute status, we have to consider changing its design because, as long as it has a two-terminal design, it can't be a constant power source.

Let's go back to the water jug analogy for a moment, but this time, let's redesign the water jug by adding an outlet valve close to the bottom of the jug so we can add water to the top of the jug at the same time we're taking water out of the bottom. Adding a spout close to the bottom of the jug makes it easy to take water out because the water's own weight forces the water out of the jug, and it makes it possible to add water and subtract water from the jug at the same time (see figure R-7).

Adding an outlet to the jug doesn't change the physics that allows the water to flow in and out of the jug. We can still add, store, and retrieve water using the jug. We simply changed the way the water cycles through the jug. We changed the mechanical structure of the jug by adding a spout. In addition, while the pressure created by the water in the jug doesn't change because we added another entrance to the jug, we increased the jug's efficiency and capacity—the amount of water it can cycle—by adding a spout at the bottom.

With the new outlet, we can maintain a constant water pressure inside the jug by adding and subtracting water from the jug at the same rate. If we redesign the deep-cycle battery so it has more than one entrance, we can increase its efficiency and overall capacity in the same way.

We can use the concept of an electrical hydraulic circuit as another analogy. An electrical hydraulic circuit is a closed water circuit or a loop made up of a water pump driven by an electric motor to produce a source of water pressure (see figure R-7A). The pump produces pressure to force or circulate the water through heating coils, where it is heated, then to the room radiators, where heat is dissipated, and back again through the water pump and heating coils. In this circuit, the water has a continuous source of pressure from the water pump.

If we redesign the deep-cycle battery so it has an entrance and an exit for electrons to travel in and out at the same time, like the electric hydraulic circuit, we wouldn't have to stop one cycling process to start the other. We'd have a constant voltage source from the charging source so the current could flow from the charging source through the battery to the load and then back to the source.

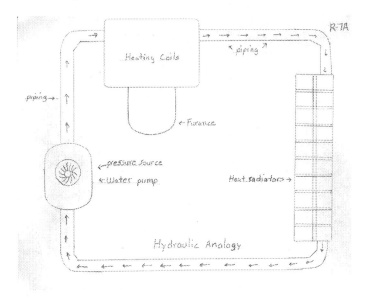

Since we can replenish the chemical energy of the deep-cycle (secondary cell) battery with electrical energy (direct current), solving the problem is simple: we must design the deep-cycle battery so we can add electrical energy back to the battery at will without interrupting its discharging process. The concept of charging the battery at will can apply to all rechargeable batteries; from a lead acid battery to a lithium ion battery, because the concept applies to the mechanical structure of the battery, not its chemical structure.

Let's assume that we designed the battery with four terminals instead of two, where all four terminals are interconnected in some form or fashion. We can create a continuous power source for the electric vehicle like that in an electric hydraulic circuit, a continuous current loop.

With a wheel-driven charging system connected to our electric vehicle's battery pack (see figure R-8), the charging system can act like the water pump in the hydraulic analogy, forcing voltage and current in and out of the battery in one direction. The batteries become the conducting source that allows the voltage and current to flow from the charging source, through the battery, to the load source, and then back to the source in one direction. Thus, the cycling functions of the battery for the electric vehicle become like an electric circuit.

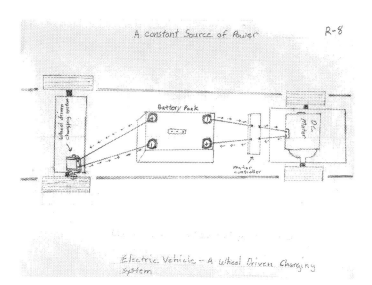

A constant Source of Power   R-8

Electric Vehicle — A wheel Driven Charging system

29

Adding electrical energy back to the deep-cycle battery at will while discharging, it does not stop the mechanics needed to discharge the battery to a load source, because we are still using an electromotive force (emf) to move electrons. In other words, we are using a magnetic emf instead of a chemical emf to move the electron current during a simultaneous cycling process or charging the battery at will. Thus, charging the battery with a higher voltage than the normal battery voltage does not only break the bond between the lead and sulfate ions but it also creates a potential difference within the battery that allows electrons to flow into the battery from the charging source. Therefore, if the battery had separate sets of terminals for charging and discharging, adding electrons back to the battery with a higher voltage than the normal battery voltage while discharging, create a potential difference between the charging and discharging terminals that allows electrons to flow to the load source as well.

CHAPTER 5

THE BATTERY'S CYCLING PROCESSES

The secondary cell battery—the deep-cycle automotive battery—is a rechargeable battery. The rechargeable batteries' chemical compositions are two-directional because we can both add and subtract electrons from their chemical composition during their cycling processes. As a result, the rechargeable battery is a power source during its discharging process and a load source during its charging process, consuming and storing energy from the power source.

Figure R-2 shows the cycling processes of the rechargeable battery with a two-terminal design in a schematic diagram that uses the electron current direction flow concept to explain the flow of the current through the battery during its cycling processes. During its discharging cycle, the current flows from the negative terminal, through the load, and back to the positive terminal of the battery; during its charge cycle, the current flows from the charging source, through the negative terminal of the battery to the positive terminal, and back to the source, the charging device. Thus, we have a complete path for the current to follow during both cycling processes: a complete circuit.

The rechargeable battery can be a power source and a load source at the same time during a simultaneous cycling process, if we redesign the deep-cycle battery so that electrons can enter and exit the battery at the same time, flowing in one direction, we can charge and discharge the battery at the same time by using magnetic emf during a simultaneous cycling process. As a result, we do not have to interrupt the current flow to the load source to recharge our batteries. Why? Because we have two loads connected in series during the simultaneous cycling process. The charging source becomes a power source for both the battery and the load at the same time.

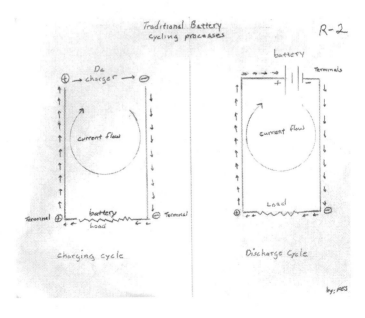

Using either the conventional current flow or the electron current flow concept, we still have a complete circuit. If you noticed, the battery becomes the load during the charge cycle, when the charger is the power source; and during the discharge cycle, the battery is the power source. Figure R-2 shows that the current flows out of the negative terminal during the discharge cycle, and it flows into the negative terminal during the charge cycle.

To come up with an idea to redesign the deep-cycle battery so both processes could be done at the same time from outside the battery, I had to understand the mechanics that take place in the lead acid battery during its cycling processes. Experts in the field of automotive battery technology say that the charge cycle of the deep-cycle (lead acid) battery is the reverse process of the discharge cycle; therefore, during the discharging cycle of the lead acid battery, electrons flow from the negative terminal (cathode), through the load source, and back to the positive terminal (anode); and during the charging cycle, electrons are forced through the battery in the opposite direction by the application of voltage across the battery's terminals (Perez, p. 33–34).

However, in order to add electrons back to the battery without interrupting the discharging process, the charging current and the discharging current must flow in one direction. Understanding these facts gave me insight into how to design the deep-cycle battery so both the charge and discharge cycle could be precipitated at the same time from outside the battery: having separate terminals for charging and discharging.

Figure R-3 shows a schematic diagram of the Miracle Auto Battery's cycling process during its simultaneous charging and discharging process, assuming that we redesign the battery with more than two terminals so that we could have a load and a charge on the battery at the same time.

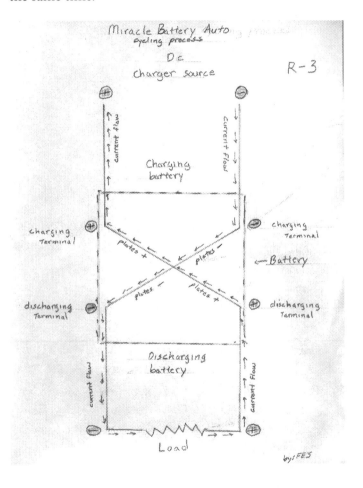

In the Miracle Auto Battery, the terminals are interconnected by way of their respective plates. In figure R-3, the line from the negative terminal crosses the line from the positive terminal over to the other negative terminal and vice versa. The crossover indicates where the two negative and the two positive terminals are interconnected by way of their respective plates. Since the current has to travel across the entire plate to charge the battery properly, we have to place the terminal connections on the opposite side of their respective plates so current can enter on one side of the plate and exit on the other going in one direction. Direct current can flow in only one direction at a time, so this design allows the current to flow across the plates in one direction: from the charging source, through the battery, across its plates and cells, and to the load source, and then back to the source. Now we have a complete circuit

The negative terminals and the positive terminals are connected in series by way of their respective plates (negative or positive) as well, even though it may not appear that way. As for the relationship between the two negative terminals and that between the two positive terminals, they are connected in series with one another. The reason for this connection is that, during the simultaneous charging and discharging process of the battery, the charging terminals will have a more negative polarity than the discharging terminals will because of the charging current. In other words, one negative terminal post will have a more negative polarity than the other negative terminal post; remember direct current flows from a more negative to a more positive point in a circuit.

The negative plates and the positive plates both create a bridge between their respective terminals (negative or positive) so the current can flow in one direction, from one negative terminal post, across its respective plates, and then to the other negative terminal connection. Thus, the current will have a complete path to follow from one negative terminal to the other. In short, the two negative terminals act as one terminal connection during a simultaneous cycling process because they are interconnected by their respective plates. Don't get this confused with the relationship between the negative and positive terminals themselves; they are in parallel connection with one another as far as the charging and discharging process is concerned. Figure T-5A shows the inside view of a potential design for the Miracle Auto Battery without the porous rubber separators. In this simple design, its plates are parallel connected to its terminals. The battery has additional tab connectors, straps, and terminal connections added to the opposite side of its plates and cells. Now, the battery has more than one way for current to travel in and out of it at the same time. The illustrations of the Miracle Auto Battery in this book may not be the actual design of the battery but just an illustration to give the reader a concept of the battery.

Like the modified water jug in figure R-7, we are just changing the mechanical structure of the battery, not the chemical structure of the battery that allows electrons to be added, stored, or retrieved from the battery during its cycling processes. Designing the battery with separate charging and discharging terminals enhances the battery's cycling processes by allowing us to add electrons back to the battery at will, thereby increasing the efficiency and overall capacity of the battery.

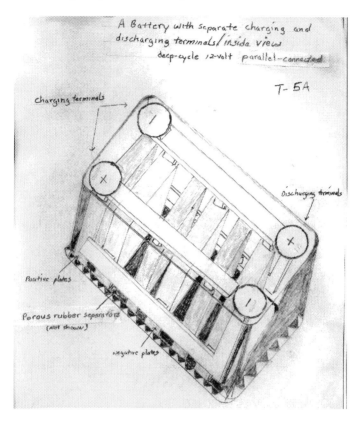

The charging current will be able to enter the battery flowing in one direction, while the discharging current will be able to exit the battery to the load source flowing in the same direction. Therefore, we can add and subtract electrons from the battery at the same time without having to stop one cycling process to start the other. The current will have a complete path from the charging source, through the battery, to the load source, and back to the charging source.

We are able to add electrons back to the battery without having to reverse the current flow through the battery's terminals. We are still charging and discharging the Miracle Auto Battery under the same principle as that of the traditional deep-cycle battery: with current flowing in and out of the battery in one direction during its cycling process. In the simultaneous charging and discharging process of the battery, we are just charging and discharging the battery in microseconds instead of using the old-fashioned way of having to stop one cycling process to start the other.

Changing the mechanical structure of the deep-cycle battery so it has more than one entrance through which current can travel doesn't prevent the battery from going through its normal cycling processes, but it does enhance its cycling processes on a micro-level so we are no longer tied to the old, drawn-out battery-charging process.

CHAPTER 6

THE MIRACLE AUTO BATTERY VS.
THE TRADITIONAL AUTO BATTERY

To understand the concept and design of the Miracle Auto Battery, we must first understand the basic design of the traditional deep-cycle battery with two terminals. If we look inside the battery to understand its cycling functions that is based on its design, we can understand that if we change the design of the deep-cycle battery, we can charge it at will. The mechanics of the electron current alone don't determine whether we can simultaneously charge and discharge the battery; the design of the battery is important as well. Figure T-1 shows a traditional deep-cycle battery with two terminals and the Miracle Auto Battery with four terminals.

Frank Earl

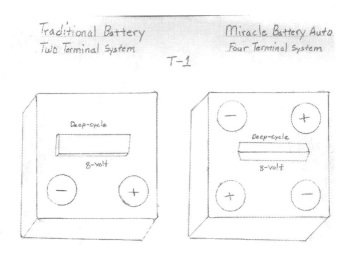

THE TRADITIONAL AUTO BATTERY

Figure T-2 shows a top inside view of the traditional deep-cycle battery with two terminals (an eight-volt battery).

The basic design of the traditional battery with two terminals in figure T-2 suggests why many think it is impossible to charge the deep-cycle battery at will and that, in any case, it wouldn't be beneficial to do so. Figure T-2 shows the basic construction of a four-cell, deep-cycle automotive battery (eight volts) with two terminals, in which each cell is separated by a plastic separator that is a part of the battery's casing. With this type of design,

40

current can pass through to each cell only by way of its straps, not by way of the electrolyte solution. Figure T-2 also shows that the cells are joined by a negative and a positive strap from each cell. This is how the current travels from one cell to the next.

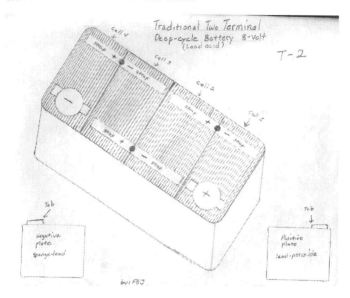

With the basic two-terminal design of the deep-cycle battery, the battery has only one path through which current can travel in or out of the battery during its charging or discharging process. The current is either going into or coming out of the battery, but it can't do both at the same time because the battery has only one entrance (two terminals) and because electrons can't pass one another going in opposite directions travelling on the same path. (Like charges repel one another.)

Figure T-4 shows the basic design of the deep-cycle battery, where each cell is composed of a number of positive

Frank Earl

and negative plates separated by porous rubber that allows the electrolyte solution to flow between the plates. This is how electrons are transferred between the positive and negative plates in each cell. In each cell, the positive plates are parallel-connected by the positive straps, and the negative plates are parallel-connected with the negative straps. The positive and negative plates are in contact with each other only by way of the electrolyte solution in each cell. With this type of design, current must pass through the negative terminal post first in order to enter or exit the battery since the positive plates are not physically connected to the negative plates. This tells us that we can use a separate set of terminals to introduce a higher voltage than the normal battery voltage into the battery to reverse the chemical reaction of the battery without having to reverse current flow through the terminal connections. Thus, we don't have to charge the battery on the same set of terminals on which it is discharged in order to recharge it.

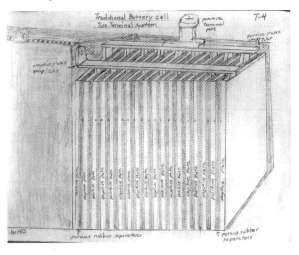

MIRACLE AUTO BATTERY

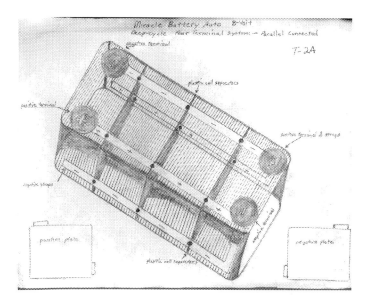

Figure T-2A shows the Miracle Auto Battery with four terminals parallel-connected. Because they are all interconnected by way of their respective plates and straps, it will be possible to add electrical energy back to the battery at will without having to interrupt the discharging process to the load source. By having two sets of terminals, two tabs per plate, and straps on both sides of the plates, we create an entrance and an exit for electrons to travel in and out of the battery in one direction (see figure T-3).

To understand the concept of the Miracle Auto Battery's simultaneous cycling functions, we can think of a 120-volt, single-phase, 60-hertz AC power system without an alternating current. If you really think about it, the battery behaves like an electric circuit to a certain degree, except it uses a chemical solution as its electromotive force during its discharging process.

Let's assume that the negative terminal and plates are the hot wires and that the positive terminal and plates are the neutral wires. Think of the discharging terminals as extra lines added to an electrical circuit panel to create dedicated lines for additional load sources. Remember that, during the charging cycle, the battery behaves like a load source that consumes and stores electrical energy. The negative plates are the power source of the battery because electrons are released from the negative plates during the bonding of the lead and the sulfate ions. Therefore, when additional straps, tab connectors, and terminals are added to the plates as discharging terminals of the battery, the plates become like an electrical circuit panel; we are just adding additional lines to the electrical box as dedicated lines for other load sources. Remember, it doesn't matter how many loads are connected to the neutral wire as long as it runs back to the source.

The reason this concept is electrically possible within the battery is that we have added additional terminals, straps, and tab connectors to the opposite side of the plates to make a two-in-one battery. Therefore, during the simultaneous cycling process, we have two loads connected in series; the battery, the load source, and the charging device become a power source for both the battery and the load source (see figure R-3).

Chapter 7

Plate Design

To achieve a one-directional current flow from the charging source, through the battery, to the load, and back to the charging source, we must focus on the plates of the battery. Since electrons have to travel from one set of terminals to the other set of terminals and across all the plates and cells of the battery at the same time, the plates must be designed so current can cross the entire plate while flowing in and out of the battery in one direction. Therefore, the plates must be designed with two tabs or lugs instead of one per plate.

Each plate has two tab connectors—one on the top and one on the bottom side of the plate (see figure S-3)—so straps can be attached to the tops and bottom sides of the plates and then to the terminal connections. With this type of design, we can create an entrance and an exit for electrons to travel across the plates and cells in one direction (see figure T-3) without having to reverse the current flow to add electrons back to the battery.

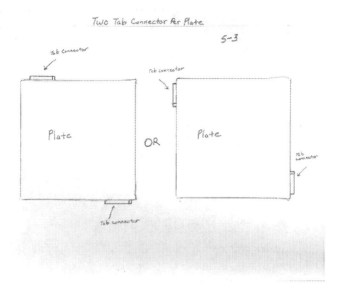

This type of plate design accomplishes two goals: First, it allows electrons to travel across the entire plate, breaking down lead sulfate while entering the battery, and to exit the battery flowing in one direction. Second, with the tab connectors on the bottom side of the plate(s), less sediment will accumulate on the straps and tabs than would be the case with the tabs and straps directly on the bottom of the plates.

We can take advantage of the mechanics of the direct current since it flows in one direction, from a more negative to a more positive point. With two tab connectors per plate, on opposite sides of the plate, we can let the current flow in and out of the battery in one direction using the voltage pressure from the charging device. Having an entrance and an exit for electrons to travel across the plates also creates a negative and a positive polarity on the same plate(s) during the simultaneous cycling process, which allows us to add and subtract electrons from the same plates simultaneously. Here's why: The terminals with the charging source will have a more negative polarity than the terminals with the load source. Because we are charging with a higher voltage than the normal battery voltage, the higher voltage pressure from the charging terminals will push the current toward the discharging terminals as the path of least resistance. Because we have a load on the discharge terminals, the current will follow the path of least resistance out of the battery to the load source. In a simultaneous cycling process, the charge and the discharge process will aid one another as well. Therefore, we don't have to reverse the current flow through the battery to recharge it.

Figure T-3 shows two plates to illustrate the cycling processes of the traditional deep-cycle (automotive) battery. Unlike the Miracle Auto Battery's plates, which have two tab connectors per plate (two entrances), the traditional deep-cycle battery has only one tab connector per plate (one entrance). Therefore, the current flow across the traditional deep-cycle battery's plates must be reversed in order to recharge the battery, because the current has only one path to enter or exit the plates. With the design of the Miracle Auto Battery's plates, current can flow across the plates in one direction while entering and exiting the battery during the simultaneous cycling process, flowing in one direction.

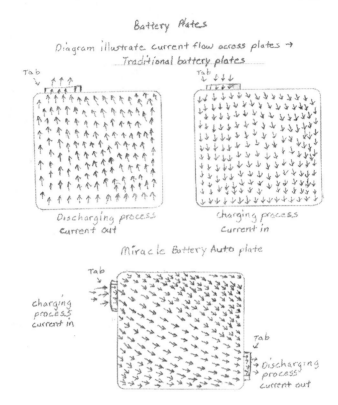

Battery Plates

Diagram illustrate current flow across plates →
Traditional battery plates

Discharging process
current out

charging process
Current in

Miracle Battery Auto plate

charging
process
current in

Discharging
process
current out

The plates of the traditional deep-cycle battery in figure T-3 show how current has to flow in one direction across the plates during the discharging process and in the opposite direction across the plates during the charging process. This type of cycling process is inefficient, time-consuming, and inconvenient because you have to stop one process to start the other.

Current follows the path of least resistance, so if we tried to add electrical energy back to the traditional battery while it had a load on it, the majority of the charging current wouldn't flow across the surface of the plates to recharge them; it would follow the path of least resistance toward the load source. The resistance within the battery would be greater than the resistance toward the load source (see figure T-3A). As a result, it is likely that not enough of the charging current would enter the battery to recharge the cells. Regardless, it wouldn't be beneficial to charge a traditional battery while it had a load on it because it would be doing nothing more than connecting two or more batteries in series or parallel to increase their overall capacity. Besides, you would only be taking energy out and not putting any back.

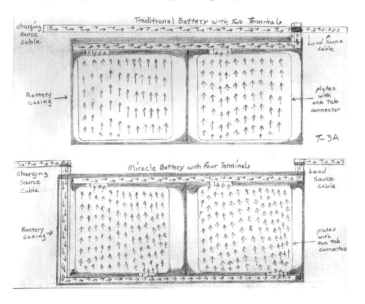

Unlike the traditional battery with two terminals, the Miracle Auto Battery has more than one set of terminals. With its unique plate designs, it allows current to flow in and out of the battery in one direction so it can add electrical energy back to the battery or charge at will without interrupting the discharge process to the load source.

Figure T-3A show that. In the Miracle Auto Battery. The charging current must cross the entire plate first because it doesn't have a direct path to the load source, as does the traditional battery with two terminals. The current enters the Miracle Auto Battery travelling across the plates and through each cell before it goes to the load source. Since the charging current must flow across the entire plate(s) first before it goes to the load source, the cells will be recharged at the same time as the charging current is entering the battery and the discharge current is exiting the battery to the load source. This process minimizes the voltage drop and the internal resistance build-up within the deep-cycle (lead acid) battery because it is charging while discharging. In other words, we are charging the battery at will.

In the traditional deep-cycle battery, electrons can only enter or exit the plate by the tab connectors from each strap in the cell. Although, electrons can travel from plate -to -plate in the cell by way of the electrolyte solution. Because the tabs and straps in each cell are on the same level (depth) in the cell, it is not beneficial to have a load and a charge on the battery at the same time. The charging current would just travel from one strap to the next in each cell until it reaches the load source, because of the resistance it encounters from the plates. Remember like charges repel one another and lead-sulfate is an insulator; thus, the charging current would follow the path of least resistance---that is, toward the load source. As a result, we get the effects of having two batteries connected in series because of the charging current is not flowing across the entire surface of the plate(s) to recharge them because of the resistance it encounters; Actually, current is flowing out of the plates along with the charging current to the load source, following the path of least resistance. Thus, we are only subtracting current from the battery during a simultaneous cycling process of the traditional deep-cycle battery. In other words, if the charging current can't cross over the surface of the plates, it can't recharge them.

Here's what I am saying, think of a forty gallon water tank with an inlet valve and an outlet valve near the bottom of the tank, same depth. You wouldn't be able to rotate the water in the tank or add and subtract water from the tank at the same time, even though; water is flowing in and out of the valves at the same time. You would only be subtracting water from the tank while having both valves open during the process, because of the relationship between the inlet and outlet valves on the tank. With both valves open, you create a path for the incoming water to travel directly out of the tank. Since the inlet and the outlet valves are at the same depth (level) in the tank, the pressure from the water that is already in the tank will be pressing down to get out and if the pressure of the incoming water is greater, the incoming water would follow the path of least resistance—this is, directly toward the open outlet valve, creating a vacuum that will draw out the water that is already in the tank along with it. Thus, the tank ends up empty during the simultaneous process. Unlike the traditional deep-cycle battery, the concept of the Miracle Auto Battery avoids this type of scenario because the charging current must enter on one side and exit on the opposite side of the plates(s), flowing across the entire surface of the plate(s) because now, the path of least resistance is travelling across the entire surface of each plate first before exiting the cells to the load source. As a result, we can recharge the plates during a simultaneous cycling process.

Chapter 8

The Logic behind the Concept

The design and cycling functions of the run capacitor help to explain why it is possible to charge the deep-cycle battery at will. The ability to add electrical energy back to the battery at will can be determined by the design of the battery, not by the mechanics of the current alone. The logic behind the concept is that we can use direct current to add and subtract electrons from the battery at the same time because of its one directional flow. In other words, If we redesign the deep-cycle battery so that it has more than one way for electrons to enter or exit it at the same time, like the run capacitor. We can use the voltage pressure from the charging source to add and subtract electrons from the battery with the mechanics of the direct current, like the alternating current does to the run capacitor in an Ac circuit, without interrupting the current flow to the load source. Thus, we can add electrical energy back to the battery at will. In addition, to adding electrical energy back to the battery at will, we can minimize lead-sulfate buildup, voltage drop, and replenish the electrolyte solution of the battery because the charging current has to cross the plates from the opposite side of the plates to exit the cells to the load source. Thus, the charging current is crossing the surface of the plates after they have been discharged to the load source. Thus, we are charging and discharging the battery in micro-seconds because we are charging from one side of the plates while electrons are flowing to the load source from the opposite side of the plates. What make this process possible to add and subtract electrons from the battery at the same time, the battery is in a charging state during the simultaneous cycling process. We are using the magnetic emf of the charging source to add and remove electrons from the battery at the same time. Remember we are not using the chemical emf of the battery to discharge it; therefore, lead-sulfate is not accumulating during the simultaneous cycling process, we are only diminishing it.

CRITICS AND SKEPTICS

Now that I have given you the rationale behind the concept and design of the Miracle Auto Battery, let's talk about the critics and skeptics for a moment. You don't have to be a rocket scientist or a genius to figure out certain things in life. You can figure them out by trial and error or by cause and effect, but most of the time, we learn from others. I compiled data from books written or supported by those in the field of automotive battery technology in order to come to a logical and scientific approach to a more efficient battery cycling process for the deep-cycle battery. The following chapters explain why it is possible to charge the battery at will and why it is beneficial to do so based on what the experts in the field of battery technology have said.

Some critics and skeptics may think it is impossible to charge the battery at will, and that in any case, there is no benefit in doing so. However, they may not have considered that we can discharge and recharge the same battery hundreds of times or that the advantage of the secondary-cell battery (rechargeable) over the primary-cell battery (non-rechargeable) is that we can replenish its chemical energy. For some reason, they have overlooked this basic fact.

Those who have a general knowledge of the cycling functions of the deep-cycle (lead acid) battery should realize that if electrical energy (dc) is used to replenish the chemical energy of the battery, then adding electrical energy back to the battery at will while discharging the battery should give us a longer-lasting battery. It should also be logical that if we can charge and discharge the same battery, then we can do these two functions simultaneously, because the active materials in the battery are designed to be able to add and subtract electrons from the same battery.

Let's us use water as an example here because we are all familiar with water. Water can be a liquid or a solid, and it can be both a liquid and a solid at the same time because they both consist of the same compounds. Scientists called it latent heat of fusion, and the layperson called it slush. Slush is where water exists as a liquid and a solid (ice) at the same time based on how much latent heat is added or removed from the water. Therefore, the amount of latent heat added or subtracted from the water determines the state of the water.

A similar scenario exists with the active materials in the deep-cycle (lead acid) battery. The compounds that we use to store electrical energy in the battery are the same compounds that allow us to retrieve electrical energy from the battery, so the chemical compounds just change state when electrons are added or subtracted during the battery's cycling processes. The electron is the catalyst in the battery's chemical composition that determines the state of the battery's chemical compounds, just as heat is the catalyst that determines the state of water. The battery's chemical compounds are in a charging or discharging state based on the number of electrons being added or subtracted.

There is a basic correlation between the number of electrons entering and exiting the lead acid battery when it comes to the lead sulfate buildup within the lead acid battery. Lead sulfate accumulates on the plates and cells of the battery during the discharge cycle, and it diminishes from the plates and cells during the charge cycle. Since the electron is the catalyst that determines the state of the battery as electrons are added or subtracted from the battery's chemical compositions, the overall capacity of the battery is based on the number of electrons entering and exiting the battery. If the conditions are right and electrons can enter and exit the battery flowing in one direction, the charging and discharging processes of the lead acid battery can occur at the same time. A battery with separate charging and discharging terminals creates the right conditions so the battery can be charged at will.

It is just a matter of which way the electrons are flowing—in or out of the battery—that determines the state of the battery's chemical composition and whether the battery is charging or discharging. Since we can replenish the chemical energy of the deep-cycle battery by applying voltage (dc) across its terminals, a logical approach is to design the battery so we can add electric energy back to the battery without interrupting the battery's discharge process to the load source. After all, scientists and engineers designed the battery so that it could be recharged with electrical energy.

What is surprising to me is that it is not obvious to skeptics that the lead acid battery is capable of doing both processes simultaneously because we use the same electron current (dc) to recharge the deep-cycle battery that we use to power the load source. In other words, we use electrical energy (dc) to replenish the chemical energy of the battery, and when we retrieve the chemical energy from the battery, it will be electrical energy (dc) again. It is only the chemical composition that changes state and the type of electromotive force that is used changes during the cycling processes of the lead acid battery, not the electron current itself. We will lose some electrical energy in the form of heat during the conversion processes, but otherwise the electron current remains the same. In short, we can still move electrons in and out of the battery with the high voltage pressure from the charging source while minimizing the lead sulfate buildup within the battery without using the chemical electromotive force of the battery. In other words, we are using electrons to move electrons. Therefore, we are just adding and subtracting electrons to the battery's chemical composition during the simultaneous cycling process. For some reason this simple logic has eluded the skeptics.

I believe that, because we can't do both the charge and discharge process simultaneously on the same set of terminals, skeptics think it is impossible to charge and discharge the battery simultaneously because they have not considered changing the mechanical structure (design) of the deep-cycle battery to do so.

CHAPTER 9

BENEFITS OF CHARGING AT WILL

If we are going to redesign the battery so it can be charged at will, there should be a benefit in doing so or there is no purpose in doing it. Since the efficiency of the deep-cycle (lead acid) battery is determined by how much energy we put in versus how much energy we take out, if we could add more energy back to the battery than we take out during a simultaneous cycling process, we could increase the efficiency and overall capacity of the battery; therefore, designing the battery so that it may be charged at will is a good idea. However, some critics and skeptics think that charging the battery at will conflicts with the laws of conservation of energy. They feel that because energy is lost during both conversion processes (the charge and the discharge process), there is no benefit in simultaneously charging and discharging the battery. Let's use what the experts say in their books to figure this one out.

Frank Earl

WHAT THE EXPERTS ARE SAYING

The factors that affect the efficiency and the overall capacity of the lead acid batteries in general are inherent in the cycling processes of lead acid batteries. Energy loss to the conversion processes is unavoidable because, in this case, we are taking energy to make energy. Logic should tell us that we could make these factors irrelevant if we could add more energy back to the battery than we take out to compensate for the energy lost during the conversion processes.

The internal resistance (lead sulfate) and the internal electrical resistance affect the amounts of energy we put in versus the amount we take from the lead acid battery during its cycling processes. These factors cause energy losses in the form of heat during the conversion processes. When we are talking about the efficiency of the lead acid battery, we are talking about the amount of energy that we put in (stored energy) during the charging cycle versus the amount of energy we take out during the discharging cycle. Remember that we will get back only about 80 percent of the energy we put in because of the energy lost during the conversion processes.

Energy lost to heat in the cycling processes of the lead acid batteries comes in two forms: The first is heat loss from the internal electrical resistance of the batteries, which is proportional to the amount of energy being cycled through the batteries, and it is irreversible. The second is heat loss from the reaction between the ions of the battery, and this loss is reversible. When the lead acid battery is being charged, the chemical reaction within the battery is exothermic—it releases heat. During the discharge cycle, the chemical reaction within the battery is endothermic—it absorbs heat.

This emission and absorption of heat is inherent in the chemical reaction between the lead and the sulfuric acid. During the charge cycle, heat is released from the chemical reaction and the internal electrical resistance of the battery. The battery temperature rises, and all this heat energy is lost; this is the primary inefficiency in the lead acid battery energy transfer process. During the discharge cycle, however, the endothermic chemical reaction balances the heat released by the internal electrical resistance within the battery during the discharge cycle. Batteries get hot when they are charged, but not when they are discharged, but energy is still lost in both conversion processes.

Therefore, the efficiency of the battery is based on the amount of energy we put in versus the amount we take out; the more energy we put in, the more energy we can get out. If we could add electrical energy back to the deep-cycle (lead acid) battery at will, the energy lost during the conversion processes would become irrelevant to the overall benefit of charging the battery at will. Also, we will minimize those factors that consume energy in the conversion processes; therefore, energy lost during the conversion processes will be negligible compared to the overall benefit of charging at will.

Here is my point: energy losses are due to the lead-sulfate buildup within the lead acid battery. Charging the battery with a higher voltage than normal battery voltage diminishes the lead-sulfate build-up within the battery that occurs as part of the process of discharging the lead acid battery. Adding electrons back to the battery while it is discharging doesn't have the same effect as it does when we are just discharging the battery or carrying out one cycling process at a time. As more of the lead-sulfate diminishes, there will be less internal resistance within the battery and less heat energy lost in the conversion processes.

Internal electrical resistance occurs when free electrons collide with fixed electrons in the elements that are being used to conduct current in the battery. Electrical resistance also occurs in our home electrical circuits, such as the copper wiring in our homes. The electrical resistance in the wiring doesn't affect the amount of energy that we can receive from our local utility company in the same way that it affects the amount of energy we receive from our batteries, because our batteries don't have a constant voltage source like our homes do. If we could supply a constant voltage source to our batteries like the utility company supplies to our homes, then the energy lost from the internal electrical resistance would be irrelevant in the conversion processes and would not affect the amount of energy needed to power our electric vehicles from point A to point B.

Since heat is absorbed during the discharge cycle of the lead acid battery, if we could add electrical energy back to the lead acid battery while it is discharging, then we could minimize the heat from the reaction in the charging process and the heat released by the internal electrical resistance. Therefore, simultaneously charging and discharging the battery can be beneficial because we can minimize those factors that affect the efficiency and overall capacity of the battery by minimizing voltage drop and internal resistance within the lead acid battery. How? By adding electrons back to the battery at will. After all, this is how we restore the chemical energy in the lead acid battery.

The consensus of experts in the field of automotive battery technology is that the charging cycle is the reverse process of the discharging cycle of the deep-cycle (lead acid) battery. In other words, during the discharging cycle, the internal resistance increases, cell voltage drops, and heat is absorbed, and the opposite happens during the charging cycle: the internal resistance decreases, cell voltage rises, and heat is released (Perez, p. 36–37). Theoretically speaking, if we could do both the charge and the discharge process simultaneously in the same battery, we could minimize these factors that rob the battery of its efficiency and overall capacity.

Since all the factors that affect the efficiency of the lead acid battery are interrelated, if we change one factor, we change them all. Therefore, adding electrons to the battery at will doesn't just minimize voltage drop; it minimizes lead sulfate build-up within the battery as well.

The lead sulfate is the bond between the lead and the sulfate ions in the lead acid battery. It is also an insulator (Perez, p. 33). As more electrons leave the battery, more lead sulfate builds up on the plates and cells of the battery, causing more resistance to the current exiting the battery and reducing the active area of the plates for more chemical reaction. Thus, energy loss occurs during the discharge cycle as well. However, the reverse happens when we are recharging the lead acid battery with a higher voltage than the normal battery voltage. Energy is lost in the form of heat because the internal resistance (lead sulfate). Is being broken down by the charging current and the charging current is fighting the electrical resistance of the battery, so the deeper the discharge is, the more time and energy it will take to recharge the battery. However, as the bonds between the lead and the sulfate ions are broken, sulfate ions are released back into the electrolyte solution, reducing the internal resistance that exists in the battery. The internal electrical resistance is proportional to the amount of energy cycled through the battery, so the energy lost because of the internal electrical resistance in the battery becomes irrelevant to the overall benefit of being able to recharge the battery at will, a cost of doing business like that of the traditional deep-cycle battery. Most of the energy loss in the form of heat is due to electrolysis, so once we minimize the internal resistance in the battery, we minimize the energy loss during the conversion processes as well. Whether we do the processes one at a time or simultaneously, energy loss will occur during the conversion processes. The question is, how much energy loss will occur?

The cycling process that we are using with the deep-cycle (lead acid) battery today is the greatest cause of the inefficiency of the lead acid battery and its inability to power the electric vehicle beyond commute status. Here's why: As we use more of its stored energy without replenishing its chemical energy right away, the battery becomes less efficient because of the ever increasing internal resistance (lead sulfate).

The deep-cycle battery is designed for deep cycling, but the paradox is that the deeper the discharge is, the less efficient the battery becomes. The whole idea of the deep-cycle battery is to get more energy out, but the more energy we take out, the harder it is to get out because of the increasing internal resistance of the battery. My point here is that, in order for the deep-cycle battery to be more efficient and a longer-lasting source of power, *we have to stop deep cycling the deep-cycle battery*. It sounds ironic, especially when the battery is built for deep cycling.

Here's why: It is harder for current to pass through a highly discharged cell than through a highly charged cell, so the deeper the discharge is, the less energy we can get from the battery and the more time and energy it will take to recharge the battery because of the highly discharged cells. Based on what experts have said about the cycling processes of the lead acid battery, there is a benefit in simultaneously charging and discharging the lead acid battery or charging it at will, because there will be less voltage drop and less internal resistance and, therefore, less energy loss in the form of heat. The result is a longer-lasting battery.

If we maintain a constant voltage level within the deep-cycle battery, as we do in the automotive starting battery, the deep-cycle battery will have less internal resistance during its conversion processes. As a result, it will take less time and energy to keep the battery charged by adding electrical energy to it at will than it would take to recharge a fully discharged battery.

ELECTRICAL BENEFITS

The electrical benefit of being able to replenish the chemical energy of the battery with electrical energy is the key to a longer-lasting battery for the electric vehicle. Remember the charge cycle is the reverse process of the discharge cycle. Designing the battery so current has to enter the battery from one side of its plates while current has to exit the battery from the other side of its plates, allows us to add electrons back to the battery at will. How? Because current is flowing to and from the plates at the same time during a simultaneous cycling process. After all, this is how we replenish the chemical energy of the secondary cell battery. If Gaston Plante' had believed that once the chemical energy of a battery has been used up and it couldn't be replenished back in 1859, we probably wouldn't have the secondary cell battery today. It is hard to believe that we can design the battery so we can replenish its chemical energy with electrical energy, but we can't design the battery so we can replenish its chemical energy while using it. That is like saying, we can recharge the battery but we can't discharge it. On the other hand, the only difference between the charging and the discharging process of the battery is the electromotive force being used to move electrons. The electron doesn't change during either cycling process; therefore, since the magnetic and the chemical electromotive force achieve the same goal of moving electrons, it should be feasible that we can add electrons back to the battery without interrupting the electron flow to the load source by using the magnetic emf of the charging source, because the power (electrons) for the load source doesn't change. In short, it is like adding gasoline to your vehicles' gas tank while the engine is running.

The whole existence of the secondary-cell battery in the first place is that, we can replenish its chemical energy with electrical energy. How? By adding electrons back to its plates, a two step process. First; the electrons must break down the resistance (lead-sulfate) on the surface of the plates; due to the chemical (emf) process of discharging the battery to a load source. Secondly, the electrons attach themselves to the surface of the plates after diminishing the lead-sulfate on the plates. The reason it takes seconds and not hours to recharge the run capacitor's plates because they have negligible resistance so the electrons can attach themselves much quicker to the plates with little resistance. My point here is that, we do not have to go through a long drawn-out charging process to recharge our batteries, if we don't allow the lead-sulfate on the plates to accumulate on a large scale from the chemical (emf) discharging process of the battery. By sending an electrical charge across the surface of the battery's plates at will, we can minimize the lead-sulfate buildup on the plates and add electrons back to the plates in minutes and not hours.

Being able to charge the batteries at will, you can extend your travel range as you drive your electric vehicle, and you will need less time and energy to add electrical energy back to your batteries because you are adding energy back as you go. Here is what I mean, let's think back to the forty gallon water tank scenario in chapter seven. If we redesign the water tank so the inlet valve is near the top and the outlet valve is near the bottom of the tank, we can actually add and subtract water from the tank at the same time. Since the incoming water has to travel the depth (from the top to the bottom) of the tank to exit through the outlet valve near the bottom of the tank, we can force water in and out of the tank at the same time with the pressure from the incoming water. If we can add and subtract the same amount of water from the tank at the same time; therefore, at any given time that we stop adding and subtracting water from the tank, we should have added back the same amount of water to the tank that we subtracted from the tank during that same period of time (adding and subtracting water).

As a result, the tank does not run empty. We will have a continuous supply of fresh water in the tank because we are rotating the water in and out of the tank. Using the concept of the Miracle Auto Battery, we can apply the same process to our deep-cycle batteries as well, so they wouldn't run empty (fully discharged) while we are using them. We can add electrons back to our batteries as we take electrons out, like the water circulating through the water tank. We can send an electrical charge from the opposite side of the plates and cells so we can reverse the chemical reaction of the battery without interrupting the electron flow to the load source. Thus, with an on board wheel driven charging system to our electric vehicles, we can increase the travel range of our vehicles on battery power as we drive them.

For example: you have a twelve volt and 100 amps battery with separate terminals for charging and discharging. The load source is a twelve volt and 20 amps device that consumes 240 watts of power per hour. With a twelve volt and 100 amps battery, the load device has five hours of run time or 1200-watt hours. We can extend the run time of the device every minute or hour it is being used by adding electrons to the battery at will. Here is why, the device is consuming .33 amps of power per minute; 20 amps divided by 60 minutes. If you are charging with a fourteen volt and 40 amp charging device and sending an electrical charge to the battery in two minutes intervals (charging process last about two minutes in duration). You are adding back .66 amps per minute; twice as much that is being consumed per minute; 40 amps divided by 60 minutes. By controlling the amount of voltage and current (.33 amps) going to the load source with variable resistors or voltage dividers, you have .66 amps across the battery per minute and only .33 amps per minute going to the load source. You have enough amps to add electrons back to the battery plus enough to keep the load source running at the same time during the simultaneous cycling process; therefore, you are adding and subtracting electrons from the battery at the same time. As a result, you can increase the device run time by adding electrons to the battery at will.

Also, we can create a very shallow cycling process by adding electrical energy back into the batteries at will. It will take less time and energy to keep the battery charged with a simultaneous cycling process due to the lower internal resistance of the battery than it would with a traditional battery cycling process, which uses about 80 percent of the batteries' capacity before attempting to recharge them. Of course, it would take more time and more energy to recharge them because the batteries' cells are highly discharged. What's more, we can recharge the batteries at a higher rate than the normal rate because we generate less internal resistance in the batteries during the simultaneous cycling process.

OTHER BENEFITS

Simultaneously charging and discharging the battery can create a very shallow cycling process with the deep-cycle battery similar to that of the automotive starting battery cycling process. A very shallow cycling process allows us to use an automotive alternator regulator control system to recharge the battery without damaging it, because the battery will never be fully discharged during the cycling processes (unless we choose to fully discharge it) (Perez, p. 103). Therefore, when the alternator throws its full load across the batteries to recharge them, it won't cook the batteries because they are not fully discharged. A simultaneous cycling process will be no different from the automotive starting battery cycling process under the alternator regulator control system because the battery will never be fully discharged.

We could also run some of the electric vehicles' other electrical components off the alternator. In addition, if we are charging with a voltage and current regulator control device—an alternator—we are using voltage dividers or variable resistors at the load source to control the amount of voltage. Using these electronic devices, we can control the amount of voltage and current entering the battery without damaging our batteries, and we can control the amount of voltage flowing to the load source without damaging the electronic devices that are on-line while we are simultaneously charging and discharging our batteries.

SUMMARY

When the lead acid battery is fully charged, there is negligible internal resistance (lead sulfate) in the battery. We know that the lead sulfate buildup in the battery is the reason for most of the energy lost during both conversion processes (charging and discharging). We also know that discharging the deep-cycle (lead acid) battery increases its internal resistance as we continue to discharge it and that charging the deep-cycle (lead acid) battery decreases its internal resistance as we continue to charge it. Therefore, simultaneously charging and discharging the deep-cycle (lead acid) battery would give us a longer-lasting source of power for electric and hybrid vehicles because we will minimize those factors that affect the efficiency and overall capacity of the battery.

The charging and discharging process of the lead acid battery can occur simultaneously because, as long as the current is entering and exiting the electrolyte solution flowing in one direction, electrons are just being added and subtracted at the same time, with some electrons lost in the form of heat during the conversion processes. Since electrons are the catalysts within the electrolyte solution that determine the state of the battery or the electrolyte solution, simultaneous charging and discharging of the battery should be feasible because we are only adding electrons to and subtracting electrons from the chemical composition of the battery.

Being able to charge and discharge the battery simultaneously at will doesn't mean that we always have to keep a constant charge on the batteries while we are discharging them, but we can add electrical energy back to the battery (charge at will) without having to stop the discharge process. In order to charge the deep-cycle battery at will, we must first consider changing the one hundred and fifty year old two-terminal design of the automotive battery.

Let's add up the facts about what experts in the field of automotive battery technology have said about the cycling processes of the lead acid battery. During the cycling processes of the lead acid battery, the internal resistance increases, cell voltage drops, and heat is absorbed in the battery during the discharge cycle, and the opposite happens during the charge cycle: internal resistance decreases, cell voltage rises, and heat is released in the battery.

Theoretically speaking, if these are the facts of the cycling functions of the lead acid battery, the benefits of being able to do both processes simultaneously is both logical and obvious. In addition, the ability to add electrical energy back to the battery at will is possible because it is a matter of which way the electrons are flowing—in or out of the battery—that determines whether the battery is being charged or discharged.

Because we can charge and discharge the same battery and that the charging cycle is the reverse process of the discharging cycle—these facts alone should raise the question; can we redesign the deep-cycle battery so we can charge it at will? Now ask yourself: Is it possible to charge the secondary-cell battery at will and receive the benefits of both cycling processes at the same time based on scientific facts? I say it is.

FINAL THOUGHTS

Crude oil was discovered in the late eighteen hundreds. Some say that it is black gold, and others say that the one who controls the oil controls those who have a demand for it. A growing demand for crude oil over the last thirty years has superseded the laws and policies that were put in place by Congress to protect some of our natural environment from oil drilling. Most of all, a growing demand for foreign oil is a threat to America's national security, which affects how we negotiate our foreign policies.

We have been using fossil fuel to power our automobiles for over a century. Today, we use hundreds of millions of barrels of crude oil every year to meet our ever-growing demand—and it is getting worse. Crude oil is not only used for gasoline that powers our automobiles; it is used for hundreds of other applications, such as synthetic materials and lubricants, and the biggest consumer of crude oil products is by far the automobile. If we could use an alternative power source to power our automobiles, we could decrease our demand for crude oil by half and end our dependence on fossil fuels to power our basic form of transportation, the automobile. This step is the key to ending our dependence on foreign oil.

Allesandro Volta made the first battery (voltaic-pile) in the late eighteenth century. Since then, we have sent men to the moon and space probes to other planets, but we have yet to produce a battery that can power the electric vehicle sufficiently to meet our primary transportation needs.

In the mid-1990s, General Motors experimented with the EV-1, an all-electric car, but the EV-1 disappeared practically overnight. Then came the hybrid cars that have both a gasoline engine and an electric motor to power the vehicle, but the car relies more heavily on the gasoline engine to power the vehicle than it does on the electric motor. The question is still asked to this day: who killed the electric car? My guess would be Big Oil, not GM, because Big Oil has the most to lose if we stop using gasoline to power our automobiles.

I am not a conspiracy theorist nor am I making a political statement. What I am saying is this: There may be more than lack of innovation involved in why we are still dependent on fossil fuels to power our automobiles. We have made great technological advancements in so many areas since the first electric cars were made in the 1890s, so why are we still using gasoline to power our automobiles? I believe it is because special-interest groups like Big Oil may be controlling the outcome of our technological advancements in the world of automotive battery technology. History has shown that we are not lacking in the area of electronic technology. We have created cell phones, iPods, computers—but we have a mysterious blind spot in the area of automotive battery technology such that the basic design of the automotive battery with its two-terminal system has not changed over the last hundred years or so. Compare that to the way capacitors have evolved over the last century.

Since we can recharge the same battery over and over again, the solution to a longer-lasting battery for the electric vehicle should be simple ... unless those who control the oil control automotive battery technology as well. The first electric cars were available over one hundred years ago; it makes you wonder why we are still dependent on gasoline to power our automobiles. Either we can design the deep-cycle battery so we can charge it at will, revolutionize the electric vehicle, and end our dependence on foreign oil; or we can keep designing the automotive deep-cycle battery the same old way for the next hundred years while our dependence on foreign oil; keep us beholden to many of the most ruthless, most despotic nations on earth.

Remember Allesandro Volta made the first primary cell battery in 1796, and then came the secondary cell battery, over fifty years later by Gaston Plante in 1859. Over 150 years have passed since the first rechargeable battery was made, and we are still using the same old cycling processes for the secondary cell battery that we used then: stopping one cycling process to start the other. It is time for us to think outside of the box (battery) to create the least expensive and most readily available clean source of energy to power our automobiles. Being able to charge and discharge the same battery hundreds of times is not an accident but by design.

I have shown you the theory behind my concept and design of the Miracle Auto Battery, a modified version of the deep-cycle (lead acid) battery. Now you can decide for yourself whether it is possible to charge the deep-cycle (lead acid) battery at will or whether the whole idea is science fiction.

Patent Pending